BEGINNER'S GUIDE TO FRESHWATER AQUARIUMS:

SETUP AND CARE MADE EASY

SIMPLE TIPS TO CREATE AND MAINTAIN
YOUR FRESHWATER AQUARIUM LIKE A PRO

Lily of the Valley

© Copyright 2023 - **All rights reserved Admore Publishing**

The content contained within this book may not be reproduced, duplicated or transmitted without direct written permission from the author or the publisher. Under no circumstances will any blame or legal responsibility be held against the publisher, or author, for any damages, reparation, or monetary loss due to the information contained within this book. Either directly or indirectly.

Contents

Foreword
An Introduction to Freshwater Aquariums

1. Getting started with your Aquarium – The Basics
    Tanks
    Substrate
    Equipment
    Filters
    Heating
    Lighting
    Plants
    Décor

2. Putting it all Together
    Location
    Substrate
    Scape the Tank
    Hardware
    Adding Water and Plants

3. Background Players: Water, Chemicals and Cycling your Tank
    Types of Water
    The Nitrogen Cycle
    Maintenance
    What NOT to Add

4. Fish
    Buying Healthy Fish
    Fish Quantity
    Fish Compatibility
    Great Beginner Fish
    Beginner to Intermediate Fish
    Clean-up Crew

5. Properly taking care of your fish
    Diet & Nutrition
    Quantity and Schedule
    Diseases, Causes, and Treatments

       Common Beginner Care Mistakes

6. Aquarist
       Joining an Aquarium Club
       Social Media and Online Groups
       Fish Shows and Contests
       Sharing your Aqua Love
       Have Fun!

Afterword
Thank You
Resources

Foreword

There is something truly magical about the fish-keeping hobby.

… But besides looking nice, does caring for fish actually add to your life?

It absolutely does! Fish tanks have so many benefits for those who own and interact with them. Studies have shown that owning fish has positive physiological and psychological effects. Having a fish tank improves your mood and reduces stress and anxiety. Observing fish elegantly float around their homes for just a few minutes lowers your heart rate and blood pressure.

There are many benefits to owning fish and maintaining an aquarium beyond this, but with all these advantages, however, comes some responsibility. Fish help us in so many ways (half of which we don't realize immediately), so it is essential that we return the favor with our care for them. This book covers everything you will need to properly care for a freshwater aquarium, and it's scaly inhabitants. This way, your flappy little guy or girl will live a very long, happy, and healthy life.

Let's get started...

An Introduction to Freshwater Aquariums

Maybe you can relate... You just bought or are about to receive your new fish friends. Excited, you rush out to make a perfect home for the newest member or members of the family. You find yourself in front of a wall of options. So many different tank sizes, so many accessories... water conditioner, PH balance, stocking options? The opportunities are endless and can become hard to manage.

All things surrounding purchasing your new fish friends can be extremely overwhelming. Researching online for the perfect environment to put your scaly finned roommates in can often just lead to more confusion. *So, it is not just as simple as putting my little guys or girls in some water and decorating it how I like?*

Unfortunately, no. But by following the simple advice in this book, it does not have to be a painstakingly tricky endeavor. Just like any other pet you have, caring for freshwater fish takes time and dedication. When I was a child and first got interested in fishkeeping, I did what many others have done and cringe to think about now. I was gifted a Betta from my parents, and I kept it in a tiny cup with no filtration or water movement. Needless to say, he sadly only lasted a few months.

The first thing any aquarist will tell you is to make sure you do your research before making ANY purchases. If I had, I would have known that Betta fish need more than just the tiny cup they come in to live long and happy lives. I would have known I needed an adequately sized home, suitable water movement, decorations, and lighting. Additionally, I would also have learned how to appropriately cycle my tank to make it safe for new inhabitants.

On my second attempt at being a Betta parent, I made sure to research the species thoroughly. I bought a five-gallon tank, driftwood, plants, heater, and substrate, then set up the tank to cycle. Only then, after cycling had completed, did I buy a beautiful blue Betta and named him Capt'n Finley. Fin lived to the ripe old age of four and a half years! Equipping myself with knowledge before getting started allowed him to live a long and happy life, which brought me great joy.

This book is meant to cover the knowledge you need to set up and run a beautiful aquarium with happy and healthy fish in it. You will gain in-depth knowledge on how to get started, set up your new tank, and how to make it a safe and comfortable place for your new pets. All this from a person who has made all the common mistakes and has learned a great deal in the process. Hopefully, you will learn from my mistakes and successes to successfully set up your first tank. Or, if you have failed before, avoid the mistakes of the past to provide a wonderful home for your new best friends.

You will also learn about different species of freshwater fish, choosing fish that will cohabitate nicely, and how to keep them healthy. Everything from water quality to décor, feeding schedules to compatibility will be covered in simple language to help you become an expert fish keeper. Then, once you can call yourself a true aquarist, I will give you some suggestions on expanding your hobby to make it even more enjoyable. By sharing your passion with others that enjoy the same interests and will even benefit from your new knowledge.

Through this book, I hope to help you build and enjoy a beautiful tank of your own with some colorful little wet friends in it. Over the years, I have researched and read through plenty of books. Still, many of them were filled with fluff, aged and no longer relevant advice, or language meant for those already having tremendous experience. Much of the information you can find online or in "freshwater aquarium" guides can be quite overwhelming, gives conflicting advice, and are often not straight to the point. This guide should be easy to read and straightforward to help you become an expert fish keeper and master of the hobby (and even the trade, if you are looking to recount your dollars from breeding fish and growing plants).

Having years of experience setting up all kinds of freshwater tanks myself, I wanted to develop an easy "how-to" guide, which leaves out all the unnecessary information you don't need in the beginning. Instead, I want to motivate and show you how easy it can be to keep happy and healthy fish, when following a few particular guidelines. If, and when, you raise your fish and plants successfully, the hobby can also become quite lucrative by selling the healthy plants and animals you raised. This is definitely not necessary, however. It's easy to create a bond with these finny little gals and guys, so you absolutely don't need to sell them if you don't want to.

This book provides the essential knowledge you need to set up a freshwater tank and fill it with life. It will give you information on which filter or heater you might want to use and which types of fish might match the aquarium style you would like to raise. However, it will not give you just this one and only path to follow. With everything in life, there are always a few options to choose from. This book will help you find the right path to create a beautiful and comfortable habitat that will give your new pets a long and happy life.

Now let us learn more about freshwater aquariums and their inhabitants …

"Aquarium fish make us realize how beautiful silence is."

— Unknown

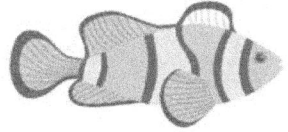

ONE

Getting started with your Aquarium - The Basics

It turns out that aquariums are much more than just a beautiful addition to any interior. They also deliver significant health benefits to their observers. They can lower people's blood pressure and heart rate, reduce stress, improve sleep quality, and even boost creativity.

While they do need a commitment of some work, aquariums are beautiful additions to any home and provide several health benefits. In this chapter, you will learn about all the basics regarding setting up your first home aquarium. We will discuss what type of tank is right for you, substrate, equipment, and plants. With this knowledge, you will be ready to start stocking up on supplies.

You can buy aquarium kits that have everything you need to get started or purchase everything separately. Kits are often a little more expensive, but they are great for beginners as there is no guesswork as to what you will need and if everything will fit together correctly. More research may need to be done if you purchase everything separately, but you can save money this way and customize the look to your heart's content!

Tanks

Size

It may seem like a good idea to start out with a very small tank. Little aquariums mean little work, right? In fact, quite the opposite is true. Small tanks get dirty extremely fast, which means you will spend more time cleaning and doing water changes than you would for a larger tank. Additionally, freshwater aquarium fish will need more room than you think. A 5-gallon tank is really only suited for a single Betta or a few freshwater shrimp. Going with a larger size will mean more options for fish and less work for you.

Until you have some fish keeping experience, stick to somewhere in the range of 10-29 gallons. However, if you are sure that fishkeeping is right for you, up to 40 gallons is not out of the question for a first tank. These sizes are a great middle ground. They are big enough that you can experiment with different décor and stocking options, but not so large to be a significant investment when first starting out. My first aquarium was 20 gallons, and I still have it stocked to this day!

Of course, the size of tank you choose also needs to fit with the animals you plan to keep. A good rule of thumb is to have at least one gallon of water for every inch of fish. For example, if you choose a 10-gallon tank, you can stock it with 10 fish that will reach one inch in length when fully grown. Alternatively, you could select 5 fish of 2 inches, or 3 fish of 3 inches, and so on.

Fish are usually tiny when first purchased, so make sure you know what size they will be when fully grown. Goldfish are typically purchased when they are very young and only about an inch in length. However, they can grow to be over a foot in length, so a small tank would not be suitable for them if you want them to live a long and healthy life. The familiar sight of goldfish in a bowl that you see in media is a MYTH! That is, unless you do not want them to live long…

When in doubt, always err on the side of caution and allow more room than you think your pets will need. All aquatic animals need enough space to swim freely and explore to keep them happy. Tanks that have too many fish will get dirtier faster. This is not only more work for you, but it also means there are more chances for your pets to get sick or even die. Happy fish show off the best colors and the most activity, so giving them ample space to live is a best practice.

Material

Next up, you will need to decide what kind of material you want the tank to be made of. The two most common are glass and acrylic. No matter what type you choose, always make sure to check for leaks before adding anything inside. This is important even if you purchase something brand new – you never know the bumps and dings your aquarium has gone through to get to you.

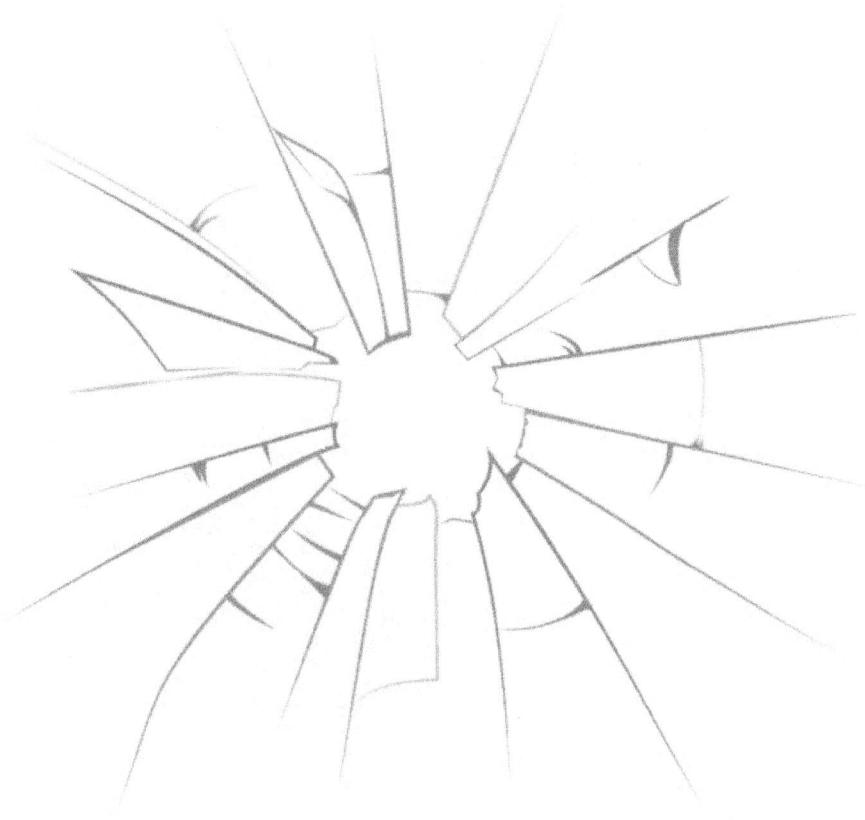

Simply fill up the tank with water and wait a few hours to make sure everything holds. If you do find a leak, simply patch it with silicone sealant, wait for it to dry, then repeat the test.

Acrylic tanks are very popular because they come in all kinds of shapes and sizes. Because they are generally made from molds, you do not have to worry about leaks due to improperly sealed corners. Being lightweight, they are easier to move around, and there is less chance of breakage when shipping or if you happen to set them down too hard or bump a table corner when moving. They are also generally less expensive than glass tanks. On the downside, acrylic scratches extremely easily, so it is essential to take extra care when cleaning your tank. Use only very soft sponges or cloths that are not abrasive enough to cause too much friction. NEVER use aquarium scrapers of any kind on acrylic tanks, or you will be sure to leave scratches.

While glass tanks are heavier and more prone to breaking, they are generally seen as higher quality than acrylic. It is much harder to scratch a glass tank when doing routine cleaning and maintenance. Most glass tanks are assembled from multiple glass pieces, so try to look for tanks made from solid pieces of glass or ensure you get a tank with high quality, undamaged silicon sealing at the seams to prevent leaks.

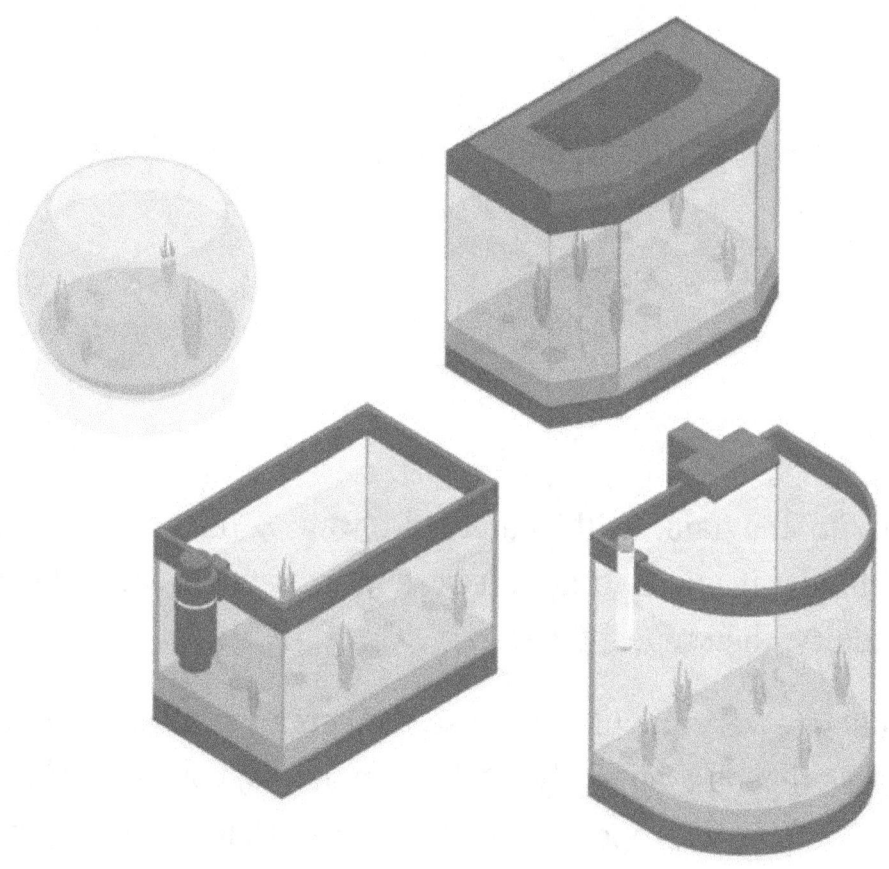

Shape

Lastly, you will need to consider tank shape. Of course, you will need to consider how much space you have when deciding where to put your aquarium. But did you know that different fish like different shaped tanks?

Some aquatic animals will need access to the surface, such as betta fish and African dwarf frogs. Because of this, they prefer homes that are shorter, so that they have easy access to the air. Others, like freshwater angelfish and discus fish, have very tall, slender bodies, so they prefer taller tanks to make them feel more at home.

Additionally, some tanks are specifically designed to produce and raise fish fry (aka, baby fish). Known as 'breeder' tanks, they are usually square and more shallow tanks than their rectangular counterparts. Making sure to choose the right tank for you, your home, and your selected aquarium inhabitants is something that requires great deliberation.

Substrate

There is a wide variety of substrates available, but which you should choose depends on your personal aesthetic and the type of fish and plants you want to keep. Be sure to also investigate what kind of fish you may want to keep. Some species can be harmed by the jagged edges of coarse rock, while others prefer it. Still, others love to roll, play, and scavenge, so soft sands would be best for them, as they will not get hurt.

I prefer natural substrate in neutral colors, such as light-colored sand or tan-colored gravel. Still, there is nothing wrong with wanting to make things more colorful! Children especially love to make things brighter with bold or even glow-in-the-dark substrates. Of course, you can always use more than one kind of substrate to create lovely contrasts. For example, one of my tanks has black gravel substrate in the back for plants, and white sand in the front to keep my cichlids happy.

You can create a river look by having dark-colored gravel on either side of the tank with a twisting strip of contrasting sand winding through it. Of course, it can be difficult having different types or colors of substrate because they can mix when vacuuming gravel or even by boisterous fish.

For a first tank, I suggest having neutral-colored, fine-grained gravel. This will satisfy nearly all beginner aquarium inhabitants. The neutral colors will make them feel as if they are in a natural environment. The fine grain will keep those with sensitive skin happy, and those that like to burrow content, to dig in the substrate to their heart's delight.

Gravel

Gravel, when it is used, should never be too coarse or largely grained. Coarse gravel can damage delicate scales and skin. Using large-grained gravel or river rocks can create nooks and crannies for fish waste and uneaten food to fall and get stuck in. This can become a breeding ground for bacteria and diseases, possibly making your fish ill and even die. So, when choosing gravel, make sure to do your homework!

Using natural colors is preferred, such as white, black, brown, and tan, because they mimic aquarium inhabitants' natural habitat. While using a colorful substrate, such as bright or florescent colors is fine, it is not a natural solution and should be avoided in most cases. Colorful substrate can be fun, especially with tanks geared toward a child's delight, but they do not occur in nature.

One concession to this rule will be if you house Glofish. Glofish are specifically bred to shine brightly under blacklight by scientifically altering species such as danios and tetras. Because they are lab-created specifically to glow in blacklight, using fluorescent, neon, or other brightly colored substrate is perfectly acceptable. They were bred specifically for these conditions, so it does not bother them at all. If you are really drawn to bright colors and will be sad if you cannot have them, use a black substrate with some bright colors mixed in. This way, both you and your fin friends will be happy!

Sand

Sand is a fantastic choice of substrate. It is soft and easily maneuverable to create interesting formations. Additionally, it is easy on delicate fins, scales, and skin. It comes in pretty much every color imaginable, so you can easily find a sand to match your aesthetic style. Another benefit of sand is that it is too fined grained for waste and uneaten food to get trapped. This makes it easy to remove uneaten food after feeding time and getting rid of fish waste when cleaning and doing water changes. Additionally, you can always have a more appropriate substrate for plant growth at the back of your tank with sand in the front. This will keep all of your plants and animals happy and healthy.

Rock

Large rocks are great for decorations, but rocks (also sometimes called river rocks) are not the best choice for substrate. As stated above, they are too large, which creates a breeding ground for bacteria and disease. However, you can use them as supplemental decorations. For example, I used small river rocks to create what looks like a cobblestone path between decorations in one of my tanks. They are placed in a single layer and pushed slightly into the surrounding fine-grained gravel to create the look. This way, there is no extra space between the gravel and the rocks for debris to become trapped, and I can still enjoy the look of river rocks. Try using my method of making a path or scattering some around the perimeter of decorations to create a more established look!

Specialized Substrate

There are several kinds of specialized substrate, such as those advertised as being for certain species of fish, specifically for shrimp, or containing nutrients that promote plant growth. While these are not necessarily essential, they usually are true to their advertising. However, you can quickly achieve the same results with specialized foods or aquarium additives. So, if the price is significantly higher than the unspecialized gravel you are considering, do not feel as if you are depriving your tank of anything essential.

Bare-Bottom

Generally, fish that prefer a bare-bottom tank belong to species that are best cared for by more advanced aquarists, such as discus fish. The reason that it is better to have no substrate for some species is because they produce excess mucus, which can foul your tank water quickly. Having substrate for these species increases the already arduous cleaning and water change schedule. However, pretty much every beginner species of fish (and plants!) will need some sort of substrate to keep them happy and healthy.

In general, when choosing a substrate, make sure the pieces are not too large, leaving lots of room between pieces. Colors found in nature are best, as they mimic aquarium inhabitants' natural surroundings to make them feel at home. Be sure to research if your fish are prone to damage from gravel or just prefer sandy bottoms, such as that of a river. No matter which you choose, as always, do your homework, and regular maintenance is key!

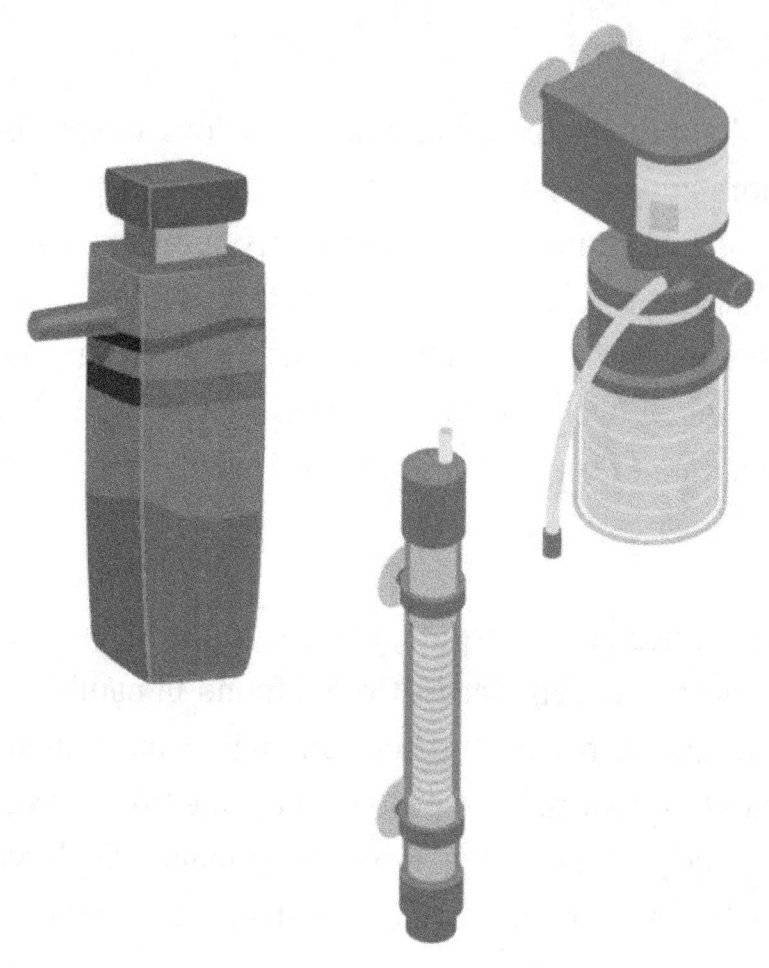

Equipment

Filters

There are lots of different types of filters to choose from. The best for beginners are ones that hang on the back of your tank (also called HOB filters, for Hang On Back). They are easy to work with, install, and are the most popular choice for medium-sized aquariums. Most hold multiple types of filter media, such as ceramics and sponges. To make sure your water is clean and creates a healthy biome for your aquarium animals.

Canister Filter

A canister filter can also be a good choice. As the name implies, all components of a canister filter are housed within a container that will sit under your aquarium in the stand's storage section. These are the most powerful filters and are better suited for larger tanks (40+ gallons). These filters are very effective, but they are often more challenging to set up and maintain, so I would not recommend them for your first few tanks.

Sponge Filter

Sponge filters are the simplest form of filter you can find. With these, water is simply pulled up a tube and then pushed back into the tank through a sponge. Sponge filters are great for smaller tanks and are routinely added to aquarium 'kits' that have everything you need to get started.

Under-gravel Filter

Not a fan of seeing all that equipment in your tank? An under-gravel filter may be for you! Under-gravel filters are placed under the substrate in your aquarium to achieve a streamlined look. These filters work great, but they take a little more maintenance and should be cleaned more often than the others on this list. Additionally, while they are great for filtering out debris that sinks down through your substrate, they are not the best for the bigger pieces of fish waste and uneaten food that rest on top of the substrate. So, you will have to regularly vacuum the top of your gravel as well. I would not recommend under-gravel filters if you plan to use sand. Sand is so fine-grained that it can clog up the mechanisms and leave you with clogged machinery.

Special Note on Filters

There is something big aquarium retailers do not want you to know ... You actually should not change your filter media often! I know, this goes against so many instincts. Still, when you change out filter and sponge cartridges, you completely remove the primary source of good bacteria in your tank!

In fact, you do not even need to buy those expensive replacement cartridges! A simple kitchen sponge (clean and never used) cut to your filter holder's size works perfectly fine. This serves to filter out any large fish waste, uneaten food, and other aquarium debris.

Then, add a small ceramic media filter, which is easily and cheaply purchased from nearly any aquarium retailer. These porous ceramics serve as a place for healthy bacteria to thrive. Place a little cheesecloth on the water output to catch any small debris, and you have everything you need!

When you do water changes, rinse all these items in the *dirty* water you are replacing. The reason for this is so you lose only a minimum of the things needed for your nitrogen cycle to thrive (*Don't worry if you don't understand things like the nitrogen cycle and bacteria yet, we will cover all of this later on in Chapter 3*). Taking these steps is not only cost-effective but will also help to maintain the health and safety of your tank. The only time you need to replace any of these items is when they start to break down. This can look like holes in cheesecloth, pieces of the sponge breaking off, or the whites of the ceramics becoming unrecognizable because of algae growth.

If any of these things happen, make sure to keep the old items in the tank for at least a week with the new items. This ensures that the good bacteria will have the time to move from the old filter media onto the new filter media. If you do not feel comfortable doing this when first starting out – fear not! Just be sure to keep your old filters concealed in the tank for at least a few days while the old filter media transfer the proper amount of good bacteria to the new.

Heating

Since most freshwater aquarium fish come from tropical regions, they generally will need a heater to match the temperatures of their native homes. Even if they can survive in lower temperatures, or if you live in a tropical or sub-tropical environment, aquarium fish will be happier with a heater. Heaters also keep your aquarium water at a stable temperature, so your fish will not be in danger of being harmed by temperature spikes.

Make sure to choose a wattage that is appropriate for your tank size. For example, a 50-watt heater would be fine for a 10-gallon tank. If you are looking more for a 40-gallon tank, a 100-300-watt heater would be more appropriate. It is always better to go for the higher end of heater wattage, as it takes less work to heat your aquarium, which will result in lower electric bills.

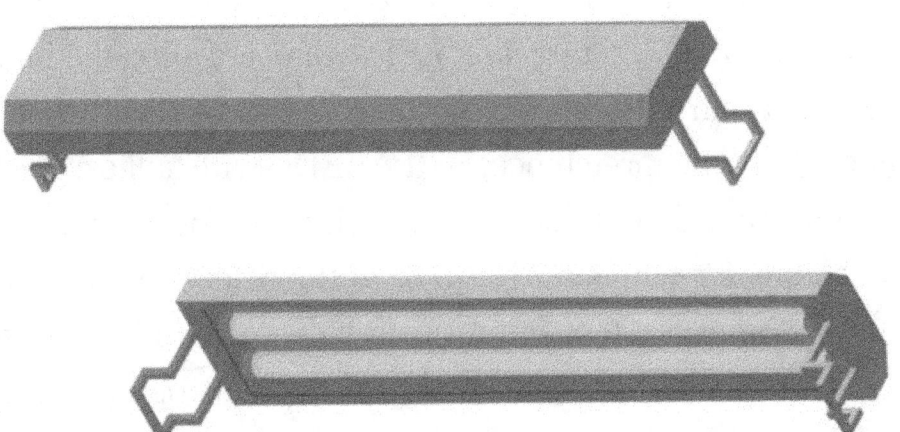

Lighting

As far as lighting is concerned, you will want to have enough light to properly see your lovely aquarium. If you decide to house live plants, you will also want to have a strong enough light for them to grow suitably. The most common form of lighting are LEDs that are already built into the underside of aquarium lids. If your lid does not have them, or they are not strong enough for your preference, extra LEDs with adhesive strips are a great option that comes very cheaply. If LEDs are not your thing, there are also lots of canopy lids with slots for t5 or t8 fluorescent bulbs that can easily be replaced. But be aware that some fin friends do not like strong lights, so providing enough plant cover and hiding places will be necessary to make them feel at home.

Plants

No one wants to stare at an empty tank all day, so now let us focus on decorations! Adding plants and décor serves a dual purpose. First, these things make your aquarium beautiful and can be really excellent ways to showcase your personality. Additionally, they make your aquarium feel like home to your new finned friends. So now you have a decision to make: artificial plants or live? Of course, you can always go with a mixture of the two or start with one and switch at a later date. Both come with benefits and drawbacks, so do your research and choose carefully!

Artificial Plants

Plants, also known as 'softscapes,' are an essential part of any fish home. Artificial plants come in various styles and colors and are very easy to take care of. When you are cleaning your tank, just give them a quick wipe down with the rest of the aquarium! Be careful when selecting artificial plants, though, as they sometimes have sharp edges that can harm the fins of more delicate fish. They are a very low-maintenance option that will still give your aquarium charm, as well as hiding places for shyer species.

Live Plants

However, I always recommend choosing live plants. They make your aquarium look like a native habitat, fish generally like them better, and it is gratifying to watch them grow! They also help to create a healthy biome in your aquarium. They oxygenate the water and eat up waste, which helps keep down the toxic bi-products that fish produce and reduce the amount of cleaning and water changes you will need to perform. As living things, they do take a little more care than plastic plants, but the extra effort is well worthwhile to me. Just make sure to be gentle with delicate leaves when removing algae and debris.

Décor

Sometimes called 'hardscapes,' decorating your tank is vital in building a safe and happy home for your fish. Some fish like to have lots of hiding places like heavy plant cover, caves, and pots. Others want lots of driftwood to explore. Still, others need lots of empty water space due to their active lifestyle. Check into what kinds of environments your fish come from and try to replicate those parameters.

I always advocate for beginners to use a mix of all kinds of decorations to find what works best for them, their tank, and their pets. For example, I have a semi-aggressive tank for my gourami's and loaches. The gourami's love to stake out their territories, so I added some large, vertical driftwood to help divide the tank into different regions. The loaches feel most secure when they have small spaces to hide in, so I created lots of small caves, nooks, and crannies by arranging large rocks and adding ceramic pots. Because it is a very large tank and I did not want to have to grow a whole underwater garden, I used a mix of live and artificial plants. With a sandy substrate, the tank resembles the fast-moving river bottoms these species call home, but in a way that looks good to me and is low maintenance. Now let us take a look at some hardscape options.

Driftwood

Driftwood makes for eye-catching scenery, and aquarium animals absolutely love it. Aquatic shrimp, such as red cherry or bamboo shrimp, will spend all day long grazing on driftwood. In addition to its aesthetic and grazing qualities, enough driftwood in your tank can actually help lower the pH levels in your water. Driftwood releases natural tannins into aquarium water, which lowers the pH and creates an environment in which it is more difficult for diseases to thrive. This can be especially helpful if you live in an area with hard water. However, lots of tannins in your aquarium will make the water look brown. Some fish like it this way! But, if you dislike the look, be sure to boil your new driftwood to reduce the tannins that will be present in your water after adding it to your tank.

Rocks

You are taking a stroll through the surf at the beach or getting ready to dive into your local lake when … OUCH! You stepped on a rock! Well, rocks are common entities in bodies of water, no matter the size. The same should be true for your aquarium. As stated before, it is best to try to create your aquarium to mimic your new pets' natural habitat. Because most regions that produce aquarium fish have some sort of rocks at the bottom, you would do well to add some too. Additionally, you can create all kinds of exciting formations for your finned friends to explore, hide in, and even stake as their own territory within the tank.

Artificial Decorations

While there is something to be said for natural beauty, can anything really beat a gorgeous castle or a ghostly sunken ship? Maybe you would like to recreate a city from Ancient Rome or a fancy Parisian shop? All of these options and so much more are available with artificial decorations. In addition to showing off your style, they also serve to make your fish feel more comfortable and secure. These decorations offer all kinds of places for your pets to explore or even hide if they are a shyer species. Just make sure you show your artificial decorations some attention when cleaning your tank to prevent algae buildup.

Before you place them in your tank, run a piece of tissue or toilet paper around any holes or openings of artificial decorations. If it can tear tissue paper, it can also tear delicate fins. If you find any rough edges, smooth them out with fine-grained sandpaper or an emery board (nail file). This will help keep all of your aquarium animals safe and secure.

No matter what kind of setup you are considering, do not forget to take personal style into consideration, though! Your aquarium should be a safe home for your pets, but it is also there for your enjoyment. There are all kinds of fun ways you can add personality to your tank and provide them a comfy atmosphere. For example, I love the world of Middle Earth created by J.R.R. Tolkien. So, in my 20-gallon tank, I added a Hobbit House decoration with a fine substrate. I added larger pebbles that lead between the house and a cave decoration to mimic the trail to a dragon's lair! Then, I found a piece of driftwood that resembled a bonsai tree and tied Christmas Moss to the branches. I finished the look by adding lots of tall, green plants in the background. Viola – a scene straight out of Lord of the Rings!

I hope this chapter has taught you about all things basic regarding setting up your first home aquarium. There are many options to choose from, but with this chapter as your guide, you will be sure to have the knowledge you need. Focus on one thing at a time: Tank, substrate, equipment, plants, and décor. Remember, you can make things simpler by buying an aquarium kit with most of the basics ready for you, or you can save a little money by purchasing everything separately. With the knowledge you now have, you are ready to start stocking up on supplies! The next thing you will need to know is how to put everything together, step by step.

"Aquariums have a place in art."
—— Unknown

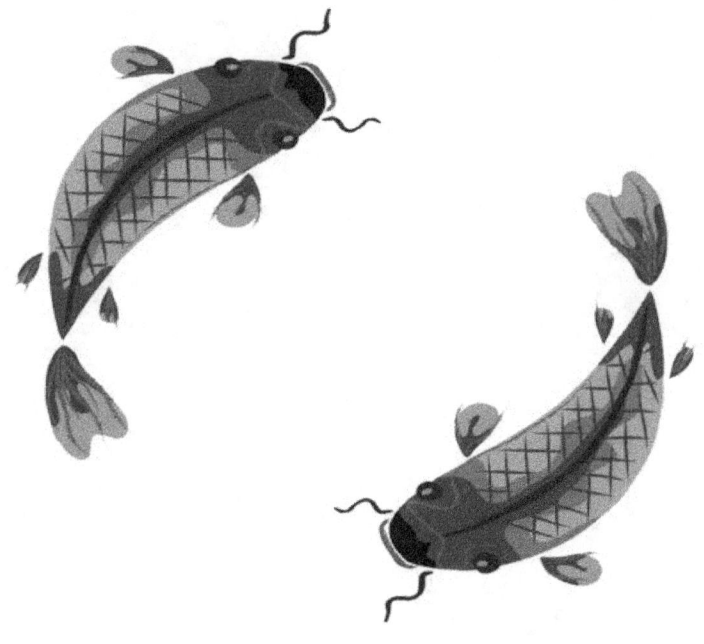

TWO

Putting it all Together

Always be yourself, unless you can be a fish, then be a fish!

Fish are extraordinary pets and awesome roommates to have. Once you own one, you may realize that these special guys and gals are not only mesmerizing to look at but also great companions. Now that you know everything you need and have figured out the essential equipment you want to use, let us put it all together into one beautiful and especially functional tank. Some things you will need to think about are where to position your tank, how to effectively place substrate, décor, filters, lighting, and how to add water without disturbing everything's arrangement.

Location

First up, you will need to find the perfect spot. You want to look for places that are not in direct sunlight or under an air vent. Aquarium animals are very sensitive to fluctuations in temperature. Being close to a window or under a vent can cause rapid changes in water temperature, which can make your pets uncomfortable or even die! If you are using a hang on the back filter, make sure you allow enough room between the tank and any walls for it to fit.

With medium to large-sized tanks (think 20+ gallons), I would suggest you invest in a quality aquarium stand. A gallon of water weighs a little over 8 pounds. Add in the substrate, rocks, and other decorations, and your tank will be hefty! A 20-gallon tank, when filled and decorated, will weigh over 160 pounds! You can save money by using solid furniture or countertops, but if you have any doubts about your current décor's ability to hold such heavy weights, invest in a wooden or metal stand.

If you have a smaller tank or are sure your tables can withstand that kind of heft, you can position your tank in the middle of a sturdy, flat surface. Make sure to clean the area thoroughly before positioning. Even small debris can cause scratches, which are not only unsightly but can also jeopardize the tank's structural integrity. There is nothing worse than waking up to find your tank has sprung a leak in the night, let alone the whole thing breaking apart. Many aquarists use a piece of foam under their tanks to prevent this from happening and ensure they do not seem to 'explode' spontaneously.

Substrate

Now that you have picked the perfect spot, it is time to start adding things to your tank! First up is the substrate. Be sure to read the packaging your substrate comes in. Some will need to be washed to remove dust and small particles that can be harmful to fish. Others contain premixed nutrients that would be destroyed with rinsing, so be sure to check which kind you have.

If you need to rinse it beforehand, the easiest way to do this is outdoors, with a five-gallon bucket and a garden hose. Put the garden hose in the bucket so that it touches the bottom, then add your substrate. This ensures that all the substrate will be properly rinsed. Turn on the water and let it run for about five minutes, oscillating the hose to make sure everything gets stirred up. Any dust or debris will be carried away with the water that runs off over the top of the bucket.

Next, add the substrate to your tank immediately because leaving damp substrate to sit around will foster the growth of bacteria. Be sure to add it carefully, so you don't accidentally chip or scratch your tank. You should have a minimum of one inch of substrate, but no more than three inches. Piling substrate too high will reduce the amount of water circulation within it, which can also be breeding grounds for bacteria.

Many aquarists like to arrange their substrate in a gentle slope, with a higher level towards the back. This creates a field of depth that is interesting to look at, and also gives more room for plants to root at the back and fish to swim in towards the front. Alternatively, you can arrange the slope from side to side or have higher levels of substrate, or hills, on which you can place decorations, rocks, and driftwood.

Scape the Tank

Now that you have your substrate arranged pleasingly and practically, it is time to start decorating! Start with your hardscape items like large rocks, decorations, and driftwood. Do not add plants yet, as they may float away when you start to add water or sustain damage while waiting for proper water levels! Make sure to allow enough room if you plan to add plants to the tank's back or sides. Play around and experiment with a few different arrangements to find the style that suits you best and creates a suitable habitat for your pets.

Hardware

Next, we will start adding your hardware equipment. Do not turn anything on yet, though, as most aquarium hardware needs to be submerged in water to avoid damage while running. Assemble and place the filter first, as it is likely the largest (and most important!) piece of equipment and will need the most room. Next, add your heater and lighting.

Adding Water and Plants

Once all your hardscape and hardware are arranged to your liking, you can start adding water to your tank. In order not to ruin the hard work you put into your decoration arrangement, you can place a small plate or bowl on your substrate. Then, pour the water onto the dish to diffuse the force of the water. As you pour, the water will then be able to trickle harmlessly over the sides, filling your tank to a level that it will not disturb your decorations.

Once your tank is about halfway full, it is time to add your plants. Whether real or artificial, it is best to place your largest pieces in the back. This creates a more in-depth look to your tank and prevents any fish or other decorations from being hidden. Now that your plants have been placed, fill the tank the rest of the way up. Leaving three to four inches of space between the top of the water and the tank lid. This will allow you to add the bags containing your new fish to the tank to acclimate (after the nitrogen cycle is complete, of course).

Now you have a gorgeous new place for your new pets to call home. You may be tempted to run to the store and start picking out some new friends – but wait! It is not yet time! You need to make sure that the water is safe for them. Next, we will look at everything you need to know about cycling your tank and achieving safe water parameters.

"If only I could live in an aquarium."

—— Unknown

THREE

Background Players: Water, Chemicals and Cycling your Tank

"Minds are like fish tanks, if you don't take care of them, they build up gunk."
– Jonathan Fader.

So, for your fish tank to not build up gunk, which can cause illness or even your fish's death, you need to put in some work and care. This chapter is all about a healthy tank with happy residents and what you need to do for general maintenance. You will learn how to treat water to make it safe for your fish, the importance of the nitrogen cycle, and how to maintain a healthy aquarium after initial set up.

Types of Water

First, you will need to fill your tank with water, but how do you know if it is safe for your fish? The key to this is knowing what your water's pH levels are and if it is in a range that will be healthy for your new friends. The best way to know the pH levels in your tap water is to purchase a freshwater aquarium testing kit. You can buy inexpensive strips for quick checks after your tank is established. Still, I highly recommend purchasing a whole testing kit. They give far more accurate results. Plus, it is a lot of fun to play chemist while measuring water into testing tubes and adding chemicals!

So, what is pH? The pH scale is a measurement of how acidic your water is. The acidity of water changes depending on the concentration of hydrogen and hydroxide ions. Acidity is linked to hydrogen ions, and bases relate to hydroxide, so adding one or the other will result in corresponding pH level fluctuations.

Pure water has a pH level of 7, which is completely neutral, neither acidic nor basic. Anything above 7 is acidic, and anything under 7 is basic. Most freshwater fish prefer water that is between 6.8 and 7.6, but it is essential to do your research. Many fish need water that is either higher or lower on the pH scale, so researching what level your fish like is vital. For beginners, look for species that prefer water on the regular scale, 6.8-7.6, as they will be the easiest to care for.

You should test your tap water at various intervals. Test it when you first put it in a bucket, then an hour later, then five hours later. This should give you a good idea of the kinds of levels your fish will be exposed to when you get to that stage of your setup. If your tap water is not acceptable, there are many ways to treat it.

Of course, you can always buy aquarium water from your local pet store, but this is often expensive and unnecessary. The tap water from most homes should be just fine when appropriately treated. If it is safe for you to drink, it will generally be safe for your tank with dechlorinating and possibly other additives. While I discourage asking for advice from big box pet stores, local businesses that specialize in freshwater fish will generally be able to give solid advice. Big box pet store employees, while competent and hardworking, usually only have basic knowledge of the animals they keep, and often little knowledge of freshwater aquarium fish. Seeking out specialized knowledge from freshwater aquarium experts is much preferred.

The Nitrogen Cycle

If you ask any experienced aquarist what their best advice is, the top answer will almost always be, "Cycle your tank!" (although, "Do your research!" would be a close second!), but what does that mean? Cycling your tank refers to waiting to add live animals to your tank until after the nitrogen cycle is complete.

So, what is the nitrogen cycle, and why does it matter? The nitrogen cycle is the process of converting ammonia into nitrates. Without going into too much biology, what basically happens this: Fish waste, dying plants, and uneaten food release ammonia into the water. High ammonia levels can be toxic to fish. Beneficial bacteria (also referred to as a 'biological filter') break down the ammonia into nitrites, then nitrates. Then, the process starts all over again, which is where the term 'cycling' comes from. Letting your tank go through the nitrogen cycle before adding any fish allows enough time for those healthy bacteria to grow and develop so they can handle the influx of ammonia when you start adding fish.

So, how do you do it? After you've completed all the steps in this book thus far – everything is decorated, and in place, your hardware equipment is all set up, and your water pH levels are satisfactory. Finish filling up your tank (if you haven't already) and ensure all equipment is running properly. Then, add a dechlorinator.

Most water sources that we humans use is treated with chlorine, chloramine, or both. This treatment is necessary to kill off pathogens and bacteria which are harmful to us. While the levels are low enough to go unnoticed by us and our furry pals (cats, dogs, hamsters, etc.), even trace amounts can be toxic to fish AND the biological filter. Adding a dechlorinator neutralizes the chlorine and chloramine to make tap water safe for aquarium inhabitants.

Next, you will need to wait for the cycle to complete. It is no fun staring at an aquarium with no fish in it for weeks (yes, I said WEEKS), so there are some ways to help speed up the process. There are three main ways to cycle your tank: fishless, fish-in, and preexisting media cycles.

<u>Fishless</u>

This type of cycling will take the longest, but also is the only way to increase your chances of raising your little fin friends without losing any, if you cannot find any preexisting media. It is often the preferred method for experienced aquarists. This cycle is basically what the name implies. There are no fish or other aquarium animals involved. You will need to buy a bacteria starter additive from your local pet store. While this is the best way, I have known other aquarists who, instead of using store-bought bacteria starters, will instead use cut up pieces of shrimp from the grocery store. Using shrimp will be a little unsightly, though, as the water tends to look cloudy until cycling is complete.

After everything is set up and running, test your water parameters daily. Within one to two weeks, you will notice the ammonia levels will spike, then gradually go back down. When the ammonia reaches one ppm (aka parts per million), add more bacteria starter or shrimp. Sometime during the second week, you will see nitrite levels rise. When they get to two ppm, stop adding bacteria starter or shrimp until it reduces to less than one ppm, then continue adding starter or shrimp. Continue this cycle until you can go at least three days without having ammonia and nitrite spikes, and you see your nitrate levels rise steadily. Once you get to this point, your tank is cycled! Do a water change of 25-50%, then you will be able to start adding the hardiest species you intend to keep. The entire process should take about four to six weeks.

Fish-In

This type of cycling is slightly faster, but you run the risk of stressing or even killing your fish. After filling your tank, you will need to start off with just three to four hardy fish, such as danios (zebra, leopard, and glofish are great!), plattys, or guppies. The reason for such a small number is so the ammonia has time to gradually build up. After they have been in the tank for two weeks, you can start slowly adding more fish. While you *can* start adding fish to your tank right after it is filled, you have less chance of losing your new fin friends if you wait at least two weeks before adding anything to the tank. This way, the healthy bacteria have a little time to start growing before the ammonia spikes.

As with the previous cycle, you will need to test your water daily to know at which point you are in during the cycling process. If at any point your ammonia levels get over one ppm, do a small water change of about 25% to avoid killing your fish. Once you see no more spikes in ammonia or nitrites and nitrate levels begin to steadily rise, you can do a partial water change and start adding more fish. It is important to add the more resilient fish first, so they can take the brunt of the stress from ammonia and nitrites while your tank goes through the cycling process. Then you can add more delicate species once your aquarium is more established.

Preexisting Media

If you have any family or friends with a well-established tank, or if you yourself have an existing tank, use some of the dirty filter material or used water from those tanks (such as sponges or ceramics) in your own filter. While it may seem odd to use dirty filter material, it is actually quite beneficial. It will cut down on the time it takes for cycling. The reason for this is because in well-established tanks, the healthy bacteria have already had plenty of time to grow and thrive, and they will happily multiply in a new tank.

This is my preferred method of aquarium cycling. It cuts down on the lengthy process and does not endanger any aquarium animals. Additionally, I have been a fish keeper for many years, so I almost always have filter media that I can move to a new tank. But, if you do not have any used filter media of your own, ask around! Most aquarists are quite happy to have someone take dirty filters off their hands and are excited to help those new to the hobby.

One piece of advice, though, is not to get preexisting media from a pet store. You never know how well the fish are taken care of, and with such a large number of rotating stock, there is a higher chance of bringing home illnesses and disease.

Notes on Cycling

Whichever method you choose to cycle your tank, patience is vital. I know it is not too much fun, but it must be done to ensure your new pets' safety and happiness. Using a water testing kit is a fun way to enjoy your tank, even when there are no fish to look at. It gives you a way to interact with your new aquarium while you wait!

Another good tip on cycling is to use live plants. It makes for a more dynamic landscape to look at, and it gives excellent clues as to how well your tank is cycling. Once you see new leaves starting to grow, it is a sure sign that your tank has finished cycling. Live plants feed on nitrates, so once nitrate levels start to rise, your plants will thrive! They also help to cut down on the number of water changes you will have to do. While nitrates are good for plants, levels that are too high can harm your fish. So, having built-in organisms that remove nitrates from the water will help to maintain a healthy cycle and cut down on work for you! While it is not strictly necessary, you may want to purchase root tabs or liquid additives to help your plants grow and look healthy.

Another good thing to note is that you may have spikes in water parameters that do not level out independently. For example, when first starting out, you could see ammonia levels rise, then keep getting higher and higher. If this happens, or any other levels continue to increase, it is crucial to do a large water change of at least 50-75%. This way, the new, clean water will dilute the unhealthy parameters and get things back on track.

Maintenance

Now that your tank is all set up, cycled properly, and your fish have been enjoying their new home, it is time to start your regular maintenance schedule. So, what should you do for routine maintenance? The answer is, it depends. Different tanks, with different sizes, shapes, inhabitants, decorations, and water parameters will vary in what kind of work they will regularly need.

For example, I have a new 5-gallon tank that houses several Blue Velvet Shrimp. That one I have to thoroughly scrub the glass and do approximately 25% water changes every week. However, with my established 55-gallon tank, I only do 25-50% water changes about once per month. I do have to vacuum it about once per week because the substrate is white sand, so fish waste and uneaten food stick out like a sore thumb. I have several species housed in it that eat algae, so I rarely have to clean the glass.

The key to proper tank maintenance when first starting out is to make sure you monitor your tank daily. Monitor the water parameters, check for algae growth, and keep an eye out for fish poop and uneaten food. The four main things you will need to focus on are partial water changes, gravel vacuuming, glass/acrylic cleaning, and filter maintenance.

Another helpful tip is that your tank should never be 'too clean.' Fish and plants need the healthy bacteria that were established in the cycling process to be happy and healthy. For the bacteria to remain happy and healthy, they need a source of food, which is provided by the fish (*see why they call it a cycle?*) So, completely emptying out your tank and scrubbing every inch of surface area is not only unnecessary but can also be very harmful – yay for less work!

Partial Water Changes

Even with a properly cycled tank, you still need to do occasional partial water changes. Think of it as if it was bathwater – would you want to use the same bathwater for months at a time? Of course not! That would get very gross, very fast. It is the same way with fish, although you probably will not have to make partial water changes as often as you bathe. Fish live full-time in their aquarium homes, and during that time, they shed mucus, eat, and poop all in the same water. While the cycling process helps cut down on waste products, it is not the same as having fresh, clean water.

Several important factors go into partial water changes. First and foremost is that they should be *partial* changes. You should never replace all the water in your tank except in the most extreme emergencies. Doing so removes most to all the beneficial bacteria that have built up in the cycling process, which means you will be back to square one. During the first month or two of setting up a new tank, you should expect to do approximately 50-75% water changes weekly or bi-weekly. As the cycling process stabilizes, you can reduce this to 25-50% changes as needed (usually bi-weekly or monthly). As stated before, this is dependent on the type of tank and fish that you have. Monitoring water levels is key!

Next, you need to ensure that the water you are replacing is treated with the same chemicals and is the same temperature as the water currently in your tank. Temperature and parameter fluctuations can be deadly to aquarium inhabitants. If your tank is room temperature, let the new water sit after running from the tap until it is also room temperature. If your tank is heated (which is recommended), use a spare heater or temporarily place your existing heater into the new water until they are roughly the same temperature. Make sure not to forget your de-chlorinator or any other water treatments you use!

Gravel Vacuuming

Even if you have the cleanest fish imaginable, you will still need to deal with fish waste and uneaten food. Letting these substances build up over time can cause harmful bacteria to thrive, which leads to sickness, diseases, and even the death of your fish. This problem is easily rectified by vacuuming your gravel. You can buy expensive, motorized vacuums that make things a little easier and less messy, but I have always used inexpensive manual vacuums. Manual vacuums use the difference in pressure between your tank water and the air to suck water up through a tube and into a bucket for disposal.

If you are using gravel, pebbles, or other rocky substrates, make sure you get the nozzle down into the gravel itself. With all those nooks and crannies, there is bound to be debris that makes its way down into the substrate layer.

This is a breeding ground for harmful bacteria, so make sure you get down past the surface layer. You do not have to agitate the entire substrate layer. Just gently push the vacuum nozzle into several high traffic areas, such as around large decorations and plant roots.

Glass/Acrylic Cleaning

Unless you have many algae eaters in your aquarium, you will need to take steps to keep algae at bay. There are several algae species, but nearly all of them are fast-growing, so it is essential to make algae cleaning a regular part of your maintenance routine. If you use acrylic tanks, use a soft sponge or cloth to gently wipe your tank. With glass, it is usually easier to remove algae with a plastic or metal aquarium scraper. Also, be sure to wipe down any big rocks or plastic decorations you have, as algae like to grow on them too!

Never use any sort of chemicals or cleaners on the inside of your tank. Even 'all-natural' options contain ingredients that will cause your pets to become ill or die. If you are having issues controlling algae in your tank, consider adding some algae-eating species! Additionally, you may also want to monitor how much food you are putting in the tank. Algae love when you overfeed, so that may be a way to cut short their food supply.

Filter Maintenance

Notice I did not say filter cleaning? The filter is where most of the good bacteria in your aquarium live and grow. It is a popular misconception that you need to change your filter or sponge materials and replace them with new ones. This concept was created by companies that sell aquarium supplies to make more money!

You can use the media that comes with aquarium kits or easily create your own. There are two things you need to create your own. First, find a brand new, never used kitchen sponge and cut it into a shape that will fit in your filter's holding area. The sponge serves to filter out pieces of debris. Next, add porous ceramics to act as a breeding ground for the good bacteria. These often come in small circular or tube shapes and are readily available at most local fish stores or online. If the holding portion of your filter is large enough, you can add these to it in addition to the sponge. If not, you can tastefully hide them around the tank or even use them as decoration!

When it is time for water changes, all you need to do is gently rinse the sponge in the used water. Take care not to rinse it in clean water, as that will remove too much of the bacteria. You do not need to scrub the sponge or even wring it out. Just a gentle rinse in the used water to remove some of the bigger pieces of gunk and debris is sufficient.

If you do decide to use pre-fit cartridges and change them out when they get dirty, make sure to put the new cartridge in the tank *with* the old one for a few days to weeks. This will allow some of the bacteria colony time to migrate over to the new cartridge. By simply tossing out the old one and replacing it, you take out the whole colony, which can be detrimental to the nitrogen cycle.

What NOT to Add

You should be extra careful when adding anything to your tank. In addition to things mentioned in previous chapters, like sharp gravel and decorations that can injure your fish, you should also be careful when adding any chemical. You may have to treat sick fish or add liquid chemicals to stabilize water parameters, but make sure to read the instruction and ingredient labels very carefully, especially with medications. Many fish medicines contain copper, which even in tiny percentages can kill certain species of fish. Do your research ahead of time to discern if your fish are sensitive to copper.

I know all this seems like a lot, but you will notice trends in timing after a few weeks. For example, you may see nitrates rise after a week – time for a partial water change! Then, you notice algae growth blooming after a month or so – time to clean the glass! The key to proper tank maintenance is to diligently monitor your aquarium at first and pay attention to trends. Before too long, you will know your tank's habits and will intuitively know when it needs attention.

Remember, your fish tank is just like your mind. If you do not take care of it, it will fill up with gunk! Now that you know how to cycle your tank and keep it well maintained, your fish are sure to be happy and healthy! Next up, we will discuss all kinds of gorgeous and entertaining aquatic friends you can look forward to purchasing!

"Aquariums tell stories that everyone interprets differently, what's your story?"

— Unknown

FOUR

Fish

After all this knowledge you have gained on choosing the right equipment and setting up your tank properly, let us finally get to the centerpiece of your home aquarium: Your new best friends! We will cover how to pick out healthy fish, how many you should purchase, and compatibility. Then we will move on to beginner-friendly aquarium inhabitants!

Buying Healthy Fish

There are several things you should look for when you go to pick out your new fin friends. The first thing you are likely to notice is their activity level. While some species are more active than others, there aren't many that are happy to just float around placidly, *especially* at the very top or bottom of the tank. Most fish are inquisitive creatures. Some are even territorial, so more often than not, you will see them exploring their surroundings, interacting with other fish, and searching for food. If you have doubts about potential fish's activity level at a store, ask to see them being fed. Even the most inactive species will perk up when it is dinner time!

The next thing to consider is coloration. Have a good idea of what the species you are interested in looks like before you get to the store. This way, if the fish for sale look dull, their fins are unhealthy, or they have any suspicious spots or discoloration, you will know that you should look elsewhere. I love doing this ahead of time, so I can plan out what kind of color schemes and activity levels I want. It is a great activity to do while your tank is cycling to pass the time!

Sometimes fish will show no outward signs of illness or disease. Even if they are showing symptoms, it can often be difficult for us humans to spot. Some common signs of illness are: protruding eyes, gasping for breath, very light or very dark spots, abnormal swimming patterns, overproduction of mucus, and fungal growth (it looks like your fish swam through thick cobwebs). Keeping an eye out for these signs will help you get the healthiest fish possible.

Fish Quantity

A common problem for first time aquarists (and sometimes experienced ones too!) is overstocking. Overstocking means that you have too many fish and not enough space for them to feel comfortable. As mentioned in Chapter 1, a good rule of thumb is to have no more than one inch of fish for every gallon of water your aquarium can hold. However, this is dependent on what kind of fish you keep and how much actual space is in your tank. If you plan to have a thick layer of substrate and a jungle-like overgrowth of plants, you may need to subtract a few to several gallons from your 1-gallon/1-inch estimate.

Another thing to ponder when considering a fish purchase is if the species you would like are schooling or shoaling fish. These types of fish will only be their happiest if they are in groups of their own kind. Whether in a store or online, any respectable fish seller will have suggestions on how many individuals of each species should be kept together. A general tip is to keep shoaling and schooling fish in groups of at least 6-10.

Fish Compatibility

Freshwater aquarium fish are usually classified as either peaceful, semi-aggressive, and aggressive. Peaceful fish are great for a community tank. They get along great with almost all other peaceful community fish. They generally will not bother or harass snails, frogs, and other fish. If peaceful fish are large enough, they may occasionally eat small aquarium shrimp or their newly hatched babies, but adult shrimp are usually too big to be considered a snack for peaceful fish. I highly recommend beginning aquarists start with a peaceful community tank. They are generally the easiest to care for, and you will not have to worry about them injuring or killing each other. Examples of peaceful fish are danios, rasboras, guppies, and mollies.

Semi-aggressive aquarium fish are usually able to be housed with their own species or other semi-aggressive fish if there are enough of them to spread the aggression around. This way, a pecking order is established where each fish knows its place in the social hierarchy, and no individual is singled out to get picked on relentlessly. Examples of semi-aggressive fish are freshwater angelfish, gouramis, and loaches.

Aggressive fish are those that will defend their territory against all other fish, even their own kind. I recommend beginners steer clear from aggressive fish unless you can keep a single individual by itself in a tank, such as a betta fish. Aggressive species generally need a high protein diet, as they are usually predators in the wild. Live, frozen, and freeze-dried meaty foods should be a staple of their diet. Examples of aggressive species are dwarf pea puffers, tiger barbs, and many species of cichlids.

Great Beginner Fish

In general, it is better to start as an aquarist with smaller or medium-sized fish. Different fish do require different types of treatment, and it would be a shame to accidentally kill some of these beautiful creatures. So, I really recommend starting off small and simple. Enjoy your little friends and as you gain more knowledge and confidence, level up slowly to more challenging species. Remember, just because they are smaller or easier to care for does not make them any less beautiful and interesting! Here is my list of great fish and other aquarium animals that are quite beginner-friendly. Find out which ones are right for you! Many aquarists like to have activity at different levels of their tank, so the following list is arranged by preferred swim levels.

Top Swimmers

Danios

Danios are some of my all-time favorite fish for all aquarists, beginner or not! They are super active little fish that are a ton of fun to watch – it is like they are almost always playing! Most varieties only grow to about two inches in length. A social fish, they like to be in groups of at least five or six. Some of the most popular danio varieties are named after the way they look, like zebra danios, leopard fin danios, and glofish (which glow under blacklight!). They get along well in almost any community tank, but their active lifestyle can be intimidating to some shyer fish. If you have any shyer fish, make sure to pay attention during feeding time so that no group of fish outcompetes others for food.

Bettas

While bettas can be found at all levels of the tank, they are put on the top swimmer's list because of their unique anatomy. They have special organs that allow them to breathe air, so access to the top of your tank is a must for them. They are stunning fish that come in tons of colors and several types of fin styles. They like to eat specialized betta pellets and small meaty foods like freeze-dried bloodworms.

Beginners should keep the more decorative male bettas on their own, though. They are very territorial, and even female bettas are known to fight each other if not kept under precise parameters. Once you have more aquarium experience under your belt, there are some other species that can be kept with bettas. Still, it's best for beginners to just have one betta in a tank on their own, with maybe a nerite snail or two to clean up uneaten food. Like goldfish, bettas are commonly seen in tiny bowls, but to keep them alive, happy, and healthy, they will need at least a five-gallon tank with a proper filter, heater, and pump. Under these conditions, they usually live from three to five years.

Middle Swimmers

Goldfish

Goldfish are probably one of the most recognizable fish in the world. They are also commonly misunderstood. Because they gained notoriety by being kept in bowls and given away as carnival prizes, it is assumed that they can live comfortably in small spaces. Nothing could be further from the truth! Goldfish actually grow up to a foot long in the wild and often reach up to ten inches in a home aquarium, so they need plenty of space. It is fine if they have smaller tanks when they are first bought, but they will need to be transferred to a bigger space as they grow.

Despite being common, they are still absolutely stunning fish. Their peaceful nature and gentle movements are sure to make you feel relaxed! They come in all kinds of colors and body shapes and are easy to feed. They will eat almost any commercially available fish flakes. They do produce a lot of waste, but if you keep up with regular tank maintenance, they are a breeze to care for!

Platies and Mollies

Platies and mollies are excellent choices for your first aquarium! They are incredibly hardy little fish that come in cool colors like red, orange, black, and my favorite, black and white spotted. They can adapt to almost any normal water parameters, so they are very forgiving fish if you make a few mistakes along the way. They are generally active swimmers and will readily accept most kinds of flaked food.

Guppies and Endler's Live Bearers

Guppies are one of the most popular freshwater species due to their vast array of colors, lively personalities, and how easy it is to breed them. Endler's livebearers come from the same family as guppies, but their bodies are usually the colorful part, and they have smaller tails. Guppies often have the most color on their long, flowy tails. Both are very easy to care for and will readily eat almost any kind of commercially available fish flakes.

Bottom Swimmers

Corydoras

Corydoras, also known as Corys or Cory Cats, are fun little mini catfish that are just so cute! They are active, curious bottom-dwellers that are always on the lookout for food. They come in various colors and stay very small, measuring in full-grown at only 1-2.5 inches. While it is possible to keep a single Cory, they are happiest in groups of 3 or more. Super adaptable and easy to care for, they are an excellent choice for a beginner community tank.

Bristlenose Plecos

While larger species of plecos will grow quite large, the Bristlenose pleco stays at a much more manageable 3-5 inches when fully grown – which is perfect for medium-sized beginner tanks. They come in colors that range from albino white to jet black with white spots, but they are mostly found in a greenish-grey hue.

Beginner to Intermediate Fish

Angelfish

Freshwater angelfish are beautiful fish that come in a variety of colors and patterns. While they are somewhat adaptable to water parameters, they need a big tank, 50+ gallons to be at their best. Because they grow to around 8 inches in height and six inches in length, they need a large, tall aquarium with lots of vertically growing plants to be their happiest. They like to have others of their own kind to socialize with but can be territorial.

Loaches

Loaches are great if you are looking for cute fish that like to hang out at the bottom of the tank. My favorites are the yo-yo loach, a black and white fish whose stripes seem to spell out 'yoyo' when they are young, and skunk loaches, who have tan bodies with a black stripe running down their spine. Different species of loaches have different personalities. In my experience, yo-yos are quite active, swimming up to the top of the tank when they know it is feeding time and eagerly exploring their environment during the day. Skunks seem to be shyer, being more active at night and preferring to take shelter in rock caves and dense vegetation. I would classify these fish as beginner to intermediate friendly, as like the angelfish, they like to have others of their kind around, but can be territorial and shy around other aggressive or semi-aggressive fish.

African Dwarf Frogs

African dwarf frogs are one of the only fully aquatic frog species. They are super fun to watch, and 2-3 frogs can be kept in as small as a 5-gallon tank. They are positively adorable little frogs that only grow to about 2 inches. However, I would classify these as perfect for beginner to intermediate aquarists. They are sensitive to water parameters and need to have easy access to the surface as they still need access to the top of the tank to get extra oxygen.

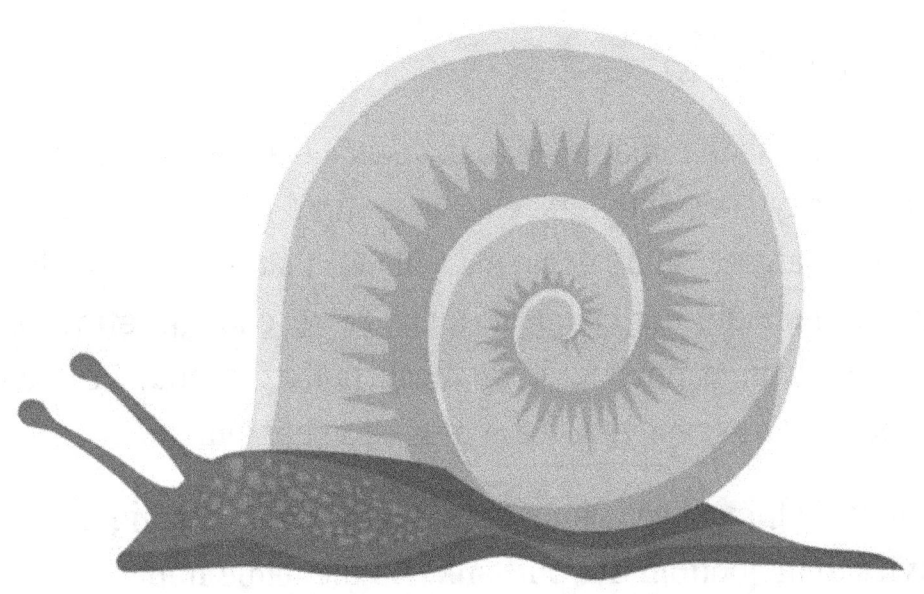

Clean-up Crew

Cherry and Ghost Shrimp

These little shrimps are remarkable to have as a clean-up crew and are tons of fun to watch. They constantly graze at the bottom of the tank to scavenge for uneaten food, algae, and decaying plant matter. Although they are scavengers, you should not count on there being enough leftovers and dead plants to keep them happy. Sinking algae pellets should be their primary source of food, with scavenging for scraps as a supplement.

In addition to helping keep your tank clean, their grazing and swimming bring activity to the bottom of your tank. As the name implies, cherry shrimp are a lovely bright red color, whereas ghost shrimp are entirely transparent! It is fascinating to watch them eat as you can see their internal organs. They both can be a little shy, so be sure to have some plant cover or caves for them to hide in. Having bolt-holes when they feel threatened will give them the courage to come out and play more often and display their best colors. You can keep them with any peaceful community fish and even semi-aggressive species if the fish are too small to consider a one-inch shrimp to be a tasty snack.

Snails

While you may be skeptical, some freshwater snails are super interesting to watch! Mystery snails come in tons of different colors and show far more activity than you would expect from a snail. They generally have solid colors like white, black, gold, purple, green, and more! Their shells are usually solid or banded, and their bodies are a lighter version of their shell color with iridescent spots. One prominent feature of the mystery snail is their long antennas, which are constantly in motion, searching for food and danger. You might also notice a tube-like structure, about half the length of their antennae, which is used to siphon water to their gills.

They are small enough (two to three inches) that you can keep one to two of them in as small as a five-gallon tank, but interesting enough to be at home in even the largest community tanks. They are completely peaceful and, like the shrimps above, are remarkable as a cleanup crew. But also, like shrimp, leftovers and dead leaves will only satisfy them between regular feedings of plant-based, sinking pellets. Happy, healthy mystery snails love to go right up to the top of the tank (secured lid necessary!), then curl up and float down to the bottom – *Whee*!

Nerite snails are another excellent option for a beginner cleanup crew that is low profile. They usually have a base color that is olive green, light to dark brown, or a dark gold, so they blend in with many naturally scaped tanks. One of the most interesting features of the nerite snail is their shell pattern. They often have whorls, zebra stripes, and dots. No two shells are the same! The only downside to the nerite is that they can lay off-white eggs all over your tank. Still, they rarely hatch, almost never survive, and are relatively easy to remove. The social and behavioral nature of nerites is much like the mystery snail, although they are slower moving.

Special Note on Snails

You should be careful about what kind of snails you introduce to your tank. Many snails can reproduce asexually, which means a single snail can produce thousands of babies if they only have enough food, space, and time. Pond snails are generally considered a nuisance and will reproduce rapidly. They do not cause much harm in small numbers, but they can overrun a tank with minimal effort and destroy plants, which can set off a chain reaction that can ruin your whole ecosystem.

Assassin snails are great for eating pond snails but can also become problematic in themselves. While both *can* be beneficial to your aquarium, for beginners, it is best to avoid them and remove them from your tank immediately if you notice them. You should take preventative measures before adding anything to your tank. But nerite and mystery snails generally do not reproduce to a nuisance level with proper tank maintenance.

These colorful jewels will be the highlight of your home decoration, but how can you ensure the health and safety of your new pets? Now that you know what to look for in your new best friends, how to pick out healthy fish, how many you should acquire, and what kind of fish will get along together, we will move on to aquarium fish care. So, how do you take care of them so they can live their best life? In the next chapter, you will learn what to feed your fish, how much and how often to feed them, Common diseases (and how to treat them), and common mistakes that will put your fish in danger of dying.

"Not all pets have fur, some have scales."

—— Unknown

FIVE

Properly taking care of your fish

"Happiness is ... having a pet fish."

So now you know how to pick out your new fin friends and what to look for regarding health, quantity, and what species will get along together. But how should you take care of these beautiful creatures? In this chapter, we will explore how to care for your new best friends properly. Happy fish show their best colors, behaviors, and activity levels, so keeping them happy and healthy will not only benefit you in being able to enjoy your tank, but will also lead them to live happily to their maximum lifespan.

Diet & Nutrition

Like humans, most beginner freshwater aquarium fish need a variety in their diet to stay at their best. Different fish need different kinds of nutrition, so which should you use for your new menagerie? The vast majority of freshwater aquarium fish are omnivorous, which means they need a diet of both protein and plant matter. The considerable amount of options in the fish food aisle can be overwhelming, and there are many choices when it comes to diets for your fish friends, so here I will break down different types of food.

Flakes

Fish flakes are definitely the most common type of food in the aquarium trade. It is an everyday staple in the diet of many aquarium inhabitants. While this is often the most cost-effective option, you should not merely look for the cheapest flakes because they differ in quality. Make sure you check out the label for the list of ingredients. The list is organized by the amount of content for each item. For example, if the first ingredient is 'shrimp,' that means that the flakes are highest in a concentration of shrimp. This is good because you know that the primary ingredient is comprised of protein. If the first ingredient is anything other than something that contains protein – stay away! But well-balanced flakes are key. The ingredient list should start with two to three protein ingredients, followed by a few plant-based ingredients. Some unpronounceable ingredients should be expected, as they are generally preservatives. Still, too many unrecognizable things should be a red flag.

Freeze-Dried Foods

Freeze-dried foods are my most preferred method of supplementing protein. As you will read below, live foods can contain pathogens and diseases, and frozen foods can cause fluctuations in temperature if not prepared properly. These foods are also great because they can be broken down into tiny pieces to fit the small mouths of many beginner freshwater fish. Both flakes and freeze-dried foods can be ground down into tiny pieces. This means that even the most miniature nano fish can get the proper diet that will make them thrive. Additionally, both flaked and freeze-dried foods will keep fresh for many months, so you can save money by buying in bulk!

Frozen Food

Many freshwater fish will snub flaked food or only eat it if they are starving. This is uncommon in most beginner fish, but it depends on the species and how they were raised. Frozen food is a fantastic option for these finicky eaters! Live foods come with risks, especially for a first-time fish keeper, so frozen foods might just be the key to your fishy's health. For those species that need a ton of protein to thrive, frozen food is optimal. It will stay fresh for several months, provides protein that your fish need, and carries little risk of having diseases, pathogens, or pests that can infect your tank. Just portion off the amount your aquarium can eat in a few days in your freezer to let it thaw, then leave it out for an hour or two to reach room temperature before adding it to your tank. Your fish are sure to devour this treat, and with proper preparation, it carries little risk.

Live Food

Some freshwater fish only thrive with live foods. But, as mentioned above, live foods do carry some risks. While it is rare, live foods such as daphnia, bloodworms, and brine shrimp can carry pathogens and diseases that can easily take root in your tank. You can do little to prevent this other than by making sure to buy your live foods from established businesses with respectable reputations. It is the best way to try to ensure disease free food.

Quantity and Schedule

It can be challenging to know how much to feed your new fish friends. They will typically continue eating for as long as they can. Some species will even eat so much that their stomachs rupture, which often causes death! Take care to only feed your fish as much as they can eat in about five minutes.

Fish are usually purchased as juveniles, and young fish need to be fed more often than when they are full-grown to ensure they reach their maximum size and best coloration. You should start out feeding them small amounts three to four times per day – no more than they can eat in around three minutes. After a few months, you can reduce the number of feedings by one. Once they have reached their full size, you can feed them one to two times per day, and as much as they can eat in five to six minutes.

Be sure to remove any uneaten food from the tank once the allotted time is up. This will prevent your fish from overeating and help keep your tank clean. Uneaten food can sink to the bottom of your tank and become trapped around plants and decorations and sink down into the nooks and crannies of substrate. This will pollute your water and creates a breeding ground for bacteria. So, do your fish friends a favor and make less work for yourself down the line by removing uneaten food!

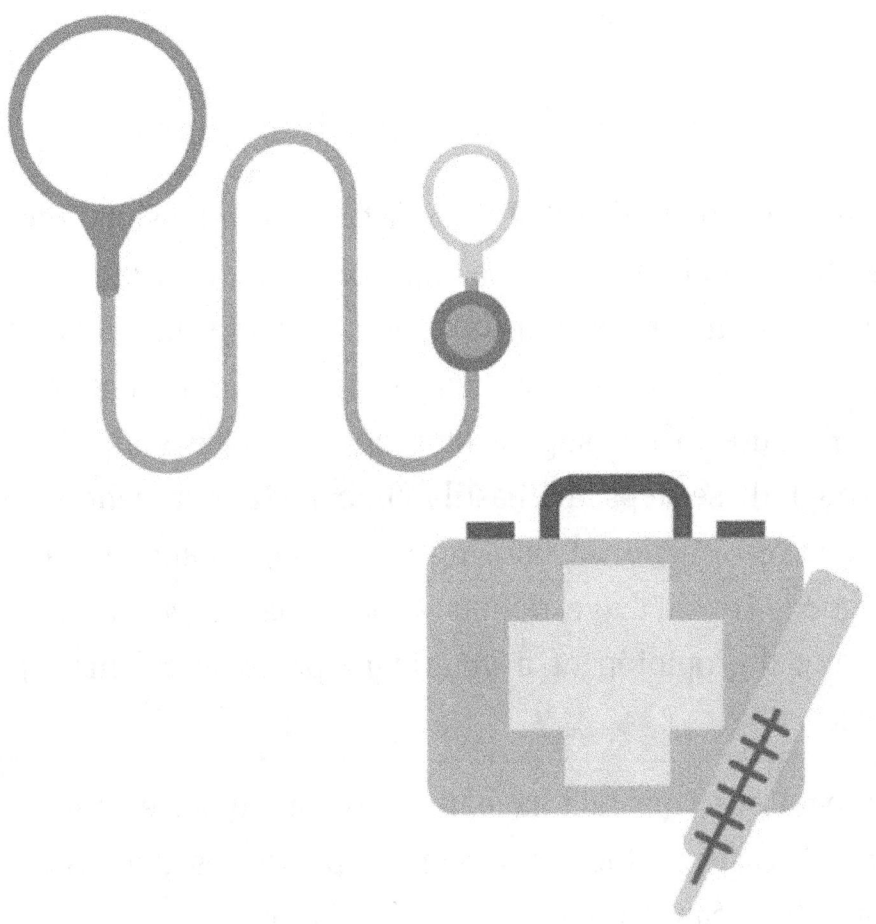

Diseases, Causes, and Treatments

What do you do if you notice that your fish seem to be sick? First, DO NOT panic. It can be very alarming to see signs of illness or stress in your beloved tank members, and it can account for you panicking into error. Take a deep breath, look closely, observe their behavior, and check your water parameters. Be sure to investigate thoroughly before you take ANY steps. Rushing through these steps and hastily choosing a treatment is definitely not the correct course of action. Choosing the wrong treatment can cause more harm than good. If you have doubts about what is wrong with your fish, please seek out the opinion of a veterinary professional that specializes in aquatic species.

No matter what issue your fish have, there is usually a treatment for common illnesses and diseases at local commercial pet stores that carry fish. Good rules of thumb for treating common illnesses is to:

1. Regularly check your water parameters. Poor water quality can not only aid in the spread of diseases but also cause them! Keeping your water clean, relatively free of uneaten food and fish waste, and adequately treated with de-chlorinator will aid in this process.

2. Use aquarium salt. Wait what?!? Salt? In a freshwater tank? The answer is, *yes*! Available in nearly all pet and aquarium stores is a small amount of aquarium salt that helps to keep illness and disease at bay. A little goes a long way in freshwater aquariums, though. Usually, just a tablespoon or so is enough for five gallons worth of water. Just be sure to follow the directions for the water to not become too salinized.

3. Use a heater to regulate the water temperature. Swings in water temperature can cause illnesses to spread and accelerate, so using a heater

with a set temperature will keep things stable.

Ick

Also called 'white spots,' Ick is one of the most common issues among freshwater aquarium fish. As the alternative name suggests, Ick is noticeable by its tell-tale white spots. It is very uncomfortable for your fish, will spread quickly to tankmates, and if left untreated or treated improperly, will cause your entire tank to fail (also known as – all your fish die!).

Ick can be introduced to your tank by buying infected fish or by maintaining improper tank parameters.

Fin Rot

Also know as fin, tail, and/or mouth rot, this disease is characterized by different parts of the fish's body looking pale, torn, or discolored. In other words, ... rotten. Fin rot usually occurs in fish that are being bullied by other fish, harassed, and being attacked. The torn fins and scales create a breeding ground for bacteria that thrive on open wounds and stressed fish.

Torn-looking Tail or Fins

Wait, did we not just cover this with fin rot? In a way. Your fish may just be getting bullied too much by other fish. If you catch it early, you can quarantine them and prevent bacteria from setting in. An excellent way to avoid a single fish from being bullied is to add more of the same species. For example, I absolutely love freshwater angelfish, but I only bought three when I first started keeping them. Two of them quickly established themselves as the 'top dogs' and paired up against the smallest one. They fin nipped mercilessly. Poor little guy. I quarantined him in his own tank until I could acquire more angelfish to help spread the aggression.

This is an example of why it is so important to research your chosen fish. Angelfish like to be in groups of at least five or six of their own kind to make sure aggression is not centered on one fish, usually the smallest one. My tank was not big enough, and I did not get enough of that species to spread aggression appropriately.

Parasites

Worms, lice, and other parasitic creatures are a sure threat to your fish. Telltale signs of these unwanted pests are visible creatures on your fish's scales, fish that seem agitated and rub themselves against plants, gravel, and other decorations, and gasping for air at the water's surface. Again, keeping appropriate water parameters will help to keep these pests to a minimum. Many of these organisms are usually present in aquarium water. Still, they will not become a problem unless you neglect regular tank maintenance.

Common Beginner Care Mistakes

Many pitfalls can trip up new aquarium owners. Listed below are some common mistakes and myths that inexperienced aquarists fall victim to. The greatest assurance against failure is education – so be sure to do your research!

The Fish Will Grow to the Tank Size

Many people have the mistaken idea, for some reason, that a fish kept in a small space will stop growing before it gets too big for its tank. I am unsure of where this mistaken idea comes from, but it is patently false. All fish species have a general adult size they will grow to, and it is essential to know how big your pets will get when fully grown to ensure you give them a proper tank size.

Over or Underfeeding

As stated above, overfeeding is an easy mistake for beginners to make. It can cause your fish to become ill or even die. But underfeeding can be just as big of a problem. Without enough nutrients in the proper proportions, your fish can also become sick and die. Even if they do not die, they will be lethargic and dull in color if they are starving.

Improper Medications

Different ailments of your fish will require different medications. It is of the utmost importance that you read the instruction label carefully and follow it exactly. Improper dosage can actually do more harm than good. If you have doubts about the kind of medication you need or the illness your fish has, be sure to consult with a veterinarian experienced with freshwater aquarium fish.

Not Cycling the Tank Properly

These next two mistakes are often the most commonly made among beginner aquarists. To make sure your tank has a healthy nitrogen cycle established is imperative to your pets' health. It takes some time for the tank to cycle, and many beginners are impatient and add fish too early. Doing so puts their lives in jeopardy.

Overstocking

Having a new tank is super exciting! You have put in so much hard work, energy, effort, and money to get everything ready, waited patiently while the tank cycled, and made absolutely sure everything was perfect for a safe and comfortable home. Now, when it is time for you to fill your tank with beautiful new life, it can be tempting to want to share your home with as many as possible. However, adding too many fish to your tank will put stress on them and create more work for you. Overstocked tanks are not a healthy environment for your fish and will substantially increase the amount of maintenance you must do to keep it clean.

Using Small Tanks or Bowls

I would never recommend using anything less than a 3-gallon tank, and that small only in cases of a single betta fish, although a 5-gallon would be preferable in that instance. Freshwater aquarium fish need space to swim and explore. Many are schooling or shoaling fish that need to be kept in medium to large-sized groups of their own kind. A tiny tank, or even worse, a bowl, is not nearly enough space to make a comfortable home for any fish. Any aquarium inhabitant will surely die prematurely if kept in such conditions.

Omitting Equipment

While some aquarium equipment may be substituted for other alternatives, omitting essential equipment such as a filter, water pump, and comfortable décor is irresponsible. Without the needed hardware, your tank will quickly become fouled and dirty. Without comfortable décor, your fish will become stressed. Without both, they will likely die.

The only way to have happy fish is to have healthy fish. Taking care of them properly will be so much more rewarding for you, as they will be active, fun, and full of color to brighten your home. Every effort you put into giving them the best home possible will be paid back tenfold. Now that you know about feeding, maintenance, and common mistakes for beginners, you have all the tools at your disposal to become a master aquarist and contribute to the fantastic community of aquarists from around the world. In the following chapter, I will relate how you can join fellow fish lovers and keepers, and both gain and impart knowledge from and to this exceptional community.

"*An aquarium is not a sport, but it's engaging for the audience.*"

—— Unknown

SIX

Aquarist

Fish lovers are united everywhere. It makes no difference where you are or the fish you keep, the community is here, and we will support you no matter what!

After all the steps you have fulfilled to set up your own tank and provide a safe and comfy home for your new pets, you can consider yourself a true aquarist. You have now joined the ranks of freshwater aquariums and fish all around the world. If you find your amazement for your new friends does not end here, you might be interested in expanding your newfound love to include others who share the same passion! You might find a great group of people right in your community, connect with others through social media, join online fish clubs around the world, or even start taking your hobby into the competitive arena with fish shows and contests!

Joining an Aquarium Club

No matter where you are located, whether in the United States like me or in the most remote regions around the world, there is most likely an aquarium club near you. Love of freshwater aquariums knows no bounds! Besides connecting you with other local freshwater fish lovers, most clubs offer things like social gatherings to meet your fellow members, fun activities, and informative newsletters and websites. If you are in the United States, an excellent place to find local freshwater fish clubs is the website *fishlab.com/local-fish-and-aquarium-clubs/*.

Have some fish that just are not what you were hoping for? Want some new aquarium inhabitants to complement your existing tank or stock a new one? Aquarium clubs can be great places to get to know other aquarists nearby that are established breeders or just looking to rotate their tank's look. Many members may even be willing to trade fish. This way, you can make sure your fin friends are going to a safe and comfortable home and, in return, you will be getting fish that are more likely to be happy and healthy. Trading fish also cuts down on expenses by allowing you to rotate your stock without paying premium prices at brick and mortar fish stores and eliminates costly shipping fees from online retailers.

No matter what you are looking for in local fish groups, joining aquarium clubs near you will be especially beneficial. You will be sure to get to know fellow fish lovers! You will also be able to have an experienced group of people who are willing to help with newcomer problems and questions. You will most likely save money by avoiding costly mistakes, as well as locate local breeders and others ready to trade fish, invertebrates, and advice for little or no cost.

Social Media and Online Groups

Not finding any physical groups locally to connect with? Or maybe you are a little iffy on in-person meetups for one reason or another? Online and social media groups dedicated to the love of freshwater aquarium fish are all over the internet. No matter where you are, social media and other online groups and clubs are at your very fingertips with nothing but an internet connection!

Personally, I like Facebook the best for these types of groups. Facebook allows you to construct questions from moderators to make sure there are no bots and that members who join are actually committed to the hobby, appropriate to the group standards, and will not post inappropriate content. It also ensures that group members comply with specific rules for posting, and most do not allow soliciting or advertising, aka asking for money, unless it is for something like a go-fund-me or other similar outside site to help raise funds for those in need of things like veterinary care. This is not to say, 'those in need of money to buy unneeded things,' but those needing extra money for vet care or to purchase equipment for sick fish and other emergency supplies.

Instagram is also a fun way to expand your freshwater aquarium love. This platform is mostly centered on photographs, so if you just want to scroll through pictures of gorgeous fish and fantastic tank setups, this is the place for you! Of course, there are many informative people to follow and knowledge to be gained. Still, the basis of this social media app is images. No matter how you use Instagram (or Insta, as it is colloquially known), you will be sure to get tons of ideas for your tank, and maybe even a little tank envy!

Pinterest is another great app. While the other sites and apps on this list are more about forging connections with other aquarists, Pinterest is more about growing your ideas with ideas from others. On this app and website, you can create personal idea boards. For example, imagine a physical corkboard or 'vision board' in your home that you can use to hold pictures secured with thumbtacks or push pins. On this 'idea board,' you place things you have cut out of magazines, made sketches of, or copied from books that you might like to include in a new or existing aquarium.

This is basically the concept of Pinterest.

But instead of creating huge, space-consuming displays in your home, it is all done from your phone or computer. Instead of spending countless days scouring magazines for pictures and articles to snip out, websites for the same and wasting money on printer ink, and/or spending hours creating drawings, everything is done in a matter of minutes to hours (however long you want to spend on it!) with the convenience of a handheld device or computer. As an added benefit, when others like your photos (also known as 'pins'), or an entire collection (also known as 'boards'), you will get a notification on who liked/saved it! This way, you can discover others who have similar boards to yours!

Best of all about Pinterest, you can have as many different boards as you like! This is a great way to work on multiple tank ideas at once. Of course, Pinterest is not only limited to freshwater aquarium ideas. You can use it to create boards for pretty much any and all interests you may have!

TikTok is relatively new to the social media scene but is growing at an extreme pace. This platform utilizes short videos to relate a simple message. What you want that message to be is entirely up to you! There are all kinds of different creators on Tiktok, and while there are not a ton of aquarists on it at this time, that does not mean that there are none or that the platform would not benefit from more! Lots of members use Tiktok to escape from reality for a few minutes (or several hours!) to watch things that are funny, relaxing, or informative in small, digestible bites. What better way to entertain and relax people than by uploading videos of stunningly colorful and relaxing videos of your tank? And the best part? You can inform people about your fascinating animals and the absolute beauty and joy this hobby provides.

Fish Shows and Contests

There are even fish shows and contests to determine the prime examples of all kinds of freshwater aquarium fish worldwide! Most contests and shows specialize in specific types of fish or generalized species umbrellas. If you can believe it, betta fish, one of the most popular beginner fish, has a thriving and highly competitive circuit in the show and contest arena.

In addition to viewing gorgeous individuals from all species, going to fish shows and contests will increase your knowledge about our amazing pets and give you tons of ideas for fish compatibility and tank setups! Once you have some experience under your belt, you can even start breeding your fin friends and entering them in contests yourself! Whether the prize is monetary or just a ribbon, you will be sure to earn some serious bragging rights among friends, family, and fellow freshwater fish lovers if you win!

Sharing your Aqua Love

We have already discussed how using social media can help expand your circle of friends and acquaintances through groups, and increase your knowledge of freshwater fishkeeping, but let us now turn to other forms of sharing your love! There is one thing I can promise you when you build your first successful tank – there will be a TON of questions. People will be curious about what you are doing and want to see your tank's progress and watch your new fin friends as they grow and play! You also will have a lot of questions yourself, so having an established group of fish lovers at your beck and call will prove to be extremely helpful.

Posting on Your Own Social Media

We have already discussed making and joining groups and following creators who love freshwater tanks, but your close friends and family are sure to be interested in your new hobby as well! Making posts about your projects and progress on your own page will draw in viewers and generate interest in your work. If you have a website, have your own online group, or video content site, one of the first things you should do to advertise is to post them on your personal Facebook, Instagram, Tumbler, Twitter, or any other site you have.

Making a Blog or Website

I am sure you know what a website is, but just in case, a website is simply an address on the internet that constitutes a page or pages with information contained within it. It usually starts with http:// or www and ends in .com, .org, .gov, and several others, depending on what type of website you are on or plan to run.

Some of you may think of a blog's definition to be laughably simple, but many may be unfamiliar with what a blog is. A blog is simply a website where you post your thoughts, feelings, and experiences on whatever subject or subjects you like! Think of it like a social media site, but rather than having to sign in to your account like you would on Facebook or Instagram, you are the only person that can log in to the account, and you are the only one who can make posts. Many website creation services allow an option for users to reply to your posts or make reviews, but you (and anyone you authorize to make posts) are the only ones that can make changes to the actual website structure.

Blogs and websites are great places to showcase your tanks and spread helpful knowledge about the freshwater aquarium hobby. Despite what you may think, website creation is relatively easy! If you can navigate the internet, email, and social media, creating a webpage will be hardly any more difficult. Companies like Google My Business and WordPress make website creation incredibly easy. If you know your way around email and social media, the process of creating a website and making posts will be no trouble at all.

Additionally, it costs very little to initially set up a website or blog, and it can even make you money in the long run. You can always monetize your site through advertising revenue. However, do not expect to get rich quick, especially at first. Advertisements on your website are a great way to earn residual income. Residual income is just a fancy term for making money from work you have already done. For instance, you create a post for your website. Then, a few months later, you receive a check or bank deposit for money generated by people clicking on the advertisements that are on your post. Easy peasy! You will not earn much, especially when you first start out, but as traffic to your site grows, so will your residual income! Be sure to promote your site on your personal social media and groups you belong to.

Making a Vlog

The main website in this day and age for vlogs is overwhelmingly YouTube. If you do not know, a 'vlog' is just a shortened term for a video blog. You can use TikTok for this type of content, but vlogs tend to be longer for videos than what TikTok will allow. Depending on your audience, videos purely for entertainment usually do best at no more than three to four minutes in length. However, if you are trying to explain a complex issue, you should take as much time as you need to fully explore the topic.

You can always make a short video on TikTok, Facebook, and/or Instagram with a link to your longer YouTube videos or videos uploaded to your blog. Always make sure to link your social media accounts and posts in the entry itself or in the comments. This way, your followers have the option to follow you on several different platforms, which not only increases their enjoyment but also increases your web visibility and advertising revenue.

Additionally, YouTube allows you to monetize your vlog channel! You know all those ads that come up before a video? Those are in place because the creator gains funds for every view and click those ads create. Once you hit one thousand followers, you will have the option to cash in on the dollars generated from advertisements. As a general rule, do not expect too much ad revenue, especially when you first start out. But every little bit helps, and once you have a large fan base, you will see your earnings rise.

Have Fun!

Talk yourself up! Tell your friends and family about your new passion. As already discussed, post about it on your social media or create a web page. You can decorate your home and workspace with freshwater fish and aquarium decorations. Buy a shirt, hat, jewelry, umbrella, or bumper sticker that features your favorite fish or freshwater aquariums in general.

The main thing about this hobby is the fun and the rewarding experience of successfully providing a safe, comfortable, and happy home for these beautiful and amazing creatures. Show enthusiasm for the hobby when you talk about it to other people. Showcase your tanks with pride in your home (or even office!) to generate interest and questions. Have tank reveal parties. Make your tanks central to your home to show off your interest and skill.

One thing I have done often is have friends and family over at feeding times. This is one of my favorite parts of being an aquarist, and it is generally well received by guests. It is so much fun to watch all your tank residents get excited, active, and playful around feeding time – some of them even make little noises if you listen close! So, sharing the beauty and activity of breakfast or dinner with your guests is sure to spread cheer.

Freshwater fish keepers come in all sizes, ages, genders, colors, and everywhere people live on earth. No matter what kind of fish you keep, what you look like, or where you are from, you are sure to find like-minded people nearby who are more than willing to share their love and knowledge of aquarium care and maintenance with you. Aquarists are generally a welcoming, friendly bunch that are sure to welcome you into the fold.

Up next, we will take a look at what we have covered so far, summarize the most important topics to refresh your memory, and motivate you (if you are not already) to get started!

"Fish are such agreeable friends. They ask no questions, they pass no criticisms."

— Unknown

Afterword

To all of those who are not familiar with fishkeeping, we can see how outsiders may think, "It's just a fish. What's the big deal?" But for those of us who love them, freshwater fish are beautiful, fascinating, and engaging pets. They each have their own personalities and have different energy levels at all different levels of the tank. While they require dedication, commitment, and regular maintenance levels, the benefits of having a fish tank in your home or office far outweigh the detriments in most cases.

So far, you have learned what type of tank is right for you to start with. Whether it be a simple five gallons setup for a single betta or a larger community tank, you now know the basics. You need a tank, filter, heater, substrate, and some decorations to make your new fin friends comfortable. Gathering all your supplies before setting it up is essential to make sure you have everything you need.

You can buy aquarium kits that have everything you need to get started or purchase everything separately. Kits are often a little more expensive, but they are great for beginners as there is no guesswork as to what you will need and if everything will fit together correctly. More research is to be done if you purchase everything separately, but you can save money this way and customize the look to your heart's content! In addition to being beautiful, they may also offer a number of health benefits to help both mental and physical health.

Once you have all of your materials, you are ready to make a space in your home and your heart for your new friends. Make sure to keep in mind what kind of tank you want, so you can place it properly in your home or office. Setting up a new tank is not easy work, but it will pay off in dividends if you do it properly. Make sure to research the fish you want to keep to determine what kind of substrate, décor, lights, and filters you will need. As described above, there is a specific order that should be maintained when adding things to your tank, so be sure to pay attention to the details in this book.

Once everything is in place, it is very tempting to run off to the pet store to immediately add new friends to your tank, but one crucial step is missing – Cycling! I know it is a pain to wait for weeks for your aquarium to go through the nitrogen cycle, but it is vital to your new pets' health. Cut down on cycling time by using existing media from established freshwater tanks your friends or family may own. Using live plants is a great way to know if the nitrogen cycle is complete. Once the plants have new leaves, it is a sure sign that your tank is safe to add new fin friends!

Next up comes maintenance, maintenance, maintenance! A buildup of uneaten food and fish poop can cause an illness, acceleration of disease, and even death. A healthy tank that contains happy residents is a tank that is clean, but not too clean. You do not have to completely empty your aquarium and scrub every inch of space. If you clean your tank too often, you will remove all the good bacteria that the nitrogen cycle needs to keep cycling. However, too much gunk will cause your fish to be unhappy, sick, and even die. Proper observation of the water parameters will let you know when maintenance is needed and when you should do water changes.

The centerpiece of your aquarium is obviously your new fin friends! But how do you choose ones that will not only look great but also live long and healthy lives? Picking out healthy fish can be tricky. You can never really know just by looking at them if individuals will show their best colors and activity levels. The most important thing here (and I am sure you are sick of hearing it) is RESEARCH!

Make sure you know what kind of water parameters your fish prefer and choose tankmates that share the same desires. Temperament is a significant consideration. You need to have fish that have the same disposition to ensure they will get along together. Then social behavior should be noted, so you know if a fish likes to be on its own, in a loose group of similar fish, or a large, tight group of their own species.

Danios, bettas, goldfish, platies, mollies, guppies, endler's, and bristlenose plecos are all excellent choices for a first-time aquarist. They are hardy, get along, and are fairly forgiving when it comes to water parameters. If you have had a tank with these types of animals before, you might try freshwater angelfish, loaches, and African dwarf frogs. While still being beginner-friendly, they may pose slightly more of a challenge.

You will need a cleanup crew in your tank, and snails and shrimp are the go-to choices for beginners. Shrimp are colorful and lively additions to any tank that will help to keep waste at a minimum. Snails too, will help to keep things clean and tidy, but both come with drawbacks.

Shrimp are a natural prey food for many freshwater fish. If you plan to breed them, do not keep them in a tank with fish that will eat their babies, which includes nearly all but the tiniest of nano fish. Even full-grown freshwater shrimp can become a tasty snack for your aquarium inhabitants. If you want to keep shrimp, keep dwarf versions with fish species that have mouths too small to consider a full-grown specimen as a meal.

Be careful also with snails. While many snail species are lovely and majestic additions to your tank, others will become a nuisance quickly. Most multiply extremely fast and can overrun your tank within a week or two. Sticking with mystery snails, which are fast-moving (for snails) and do not reproduce asexually, is a good rule to follow.

A proper maintenance and feeding schedule is essential. Know your fish, know when and how much they should be fed and follow it strictly! If you have a busy schedule and cannot be there at certain feeding times throughout the day, enlist a friend's aid or pay an acquaintance to fill in for your caretaker duties. Fish that are happy and healthy show their best colors and activity, bringing you the most joy. So, pay back the happiness you receive by making sure they have the best!

So now you have a beautiful freshwater aquarium. Your fish are happy, healthy, and comfortable. It may seem like this is the end of the road – but there is so much more! Your aquarium love can be shared with the whole world! You can post on traditional social media. You can make a website. You can create a blog or vlog that will not only grow your viewership, but also bring in an income!

Make the most of your freshwater aquarium. Take the time, give the care, show the love. I promise you it will repay you tenfold. Every ounce of energy you put into making a fantastic home for your new friends will create a lovely, relaxing, and beautiful environment for you as well.

"Aquariums light up a dark universe by giving it beauty and life."

— Unknown

Thank You

Owning an aquarium is an incredibly enriching experience. They mesmerize viewers with their gorgeous inhabitants and captivate with their tranquility. Thank you for *diving* into this fantastic hobby and learning more about taking care of these funny guys and girls. Enjoy taking care of your fish roommates, and I wish you much success and joy.

Resources

Resources for further research and information can be found below.

Aquascaping Supply. (2019). *Tank talk: The advantages of driftwood.*: https://aquascapingsupply.com/tank-talk/f/the-advantages-of-driftwood

FishLab. (2019). *Local fish and aquarium clubs.*: https://fishlab.com/local-fish-and-aquarium-clubs/

Freshwater Central. (2020). *How to cycle an aquarium.*: https://freshwatercentral.com/how-to-cycle-an-aquarium

Kenton, Will. (2020). *Corporate finance & accounting > Financial analysis: Residual income.*: https://www.investopedia.com/terms/r/residual income.asp

Pet Helpful. (2019). *Choosing healthy fish.*: https://pethelpful.com/fish-aquariums/Choosing-Healthy-Fish

Tetra Fish. (2020). *Learning center: Fish illnesses and how to spot them.*: http://www.tetra-fish.com/learning-center/troubleshooting/fish-illnesses-how-to-spot-them.aspx

Woods, R. (2019). *Best aquarium filter: Complete guide 2020.*: https://www.fishkeepingworld.com/best-aquarium-filter/

www.ingramcontent.com/pod-product-compliance
Lightning Source LLC
Chambersburg PA
CBHW081401070526

44583CB00020B/2633